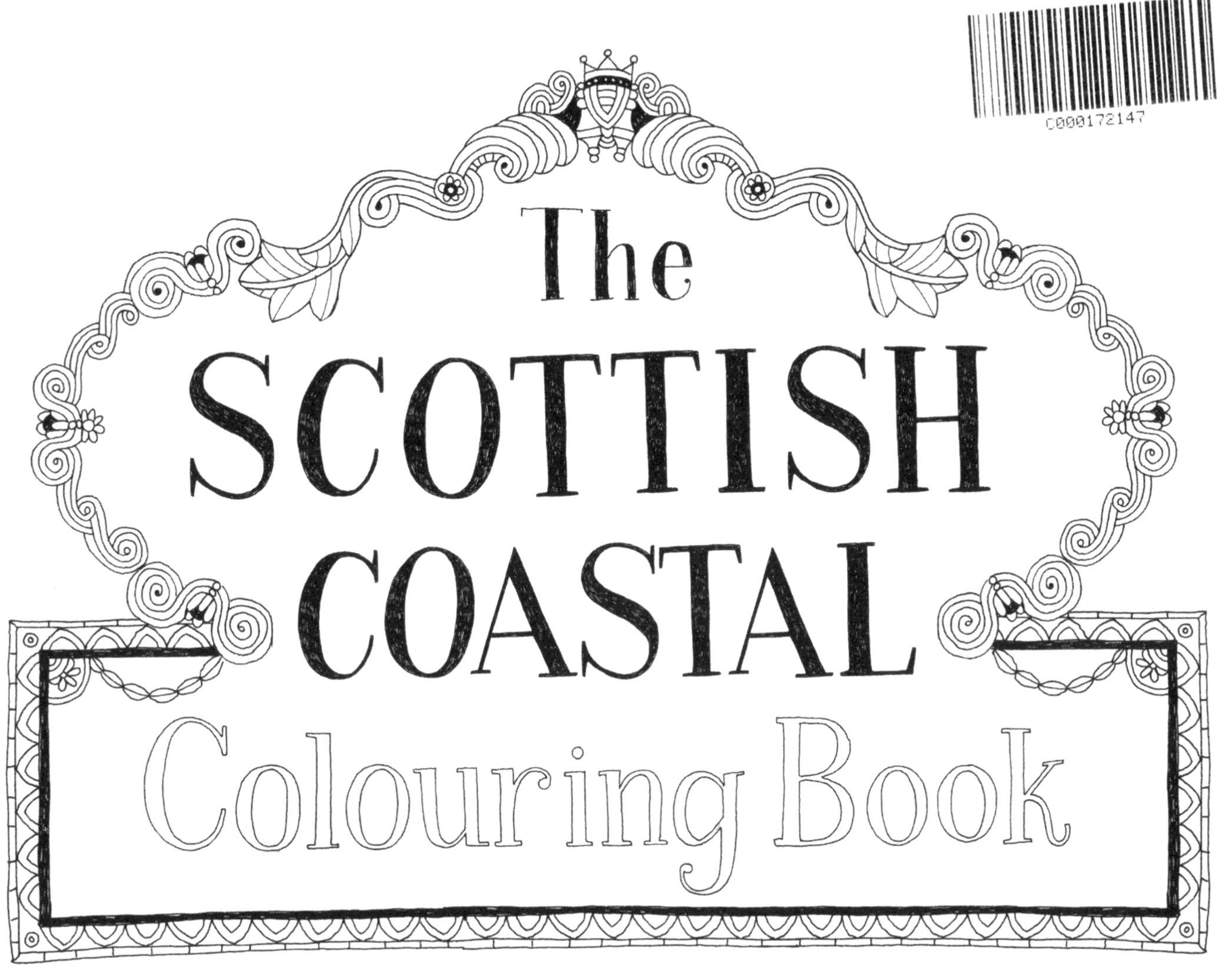


by
Eilidh Muldoon

BIRLINN

# St Abbs

# Tantallon Castle

Bass Rock

Culross

# Crail Harbour

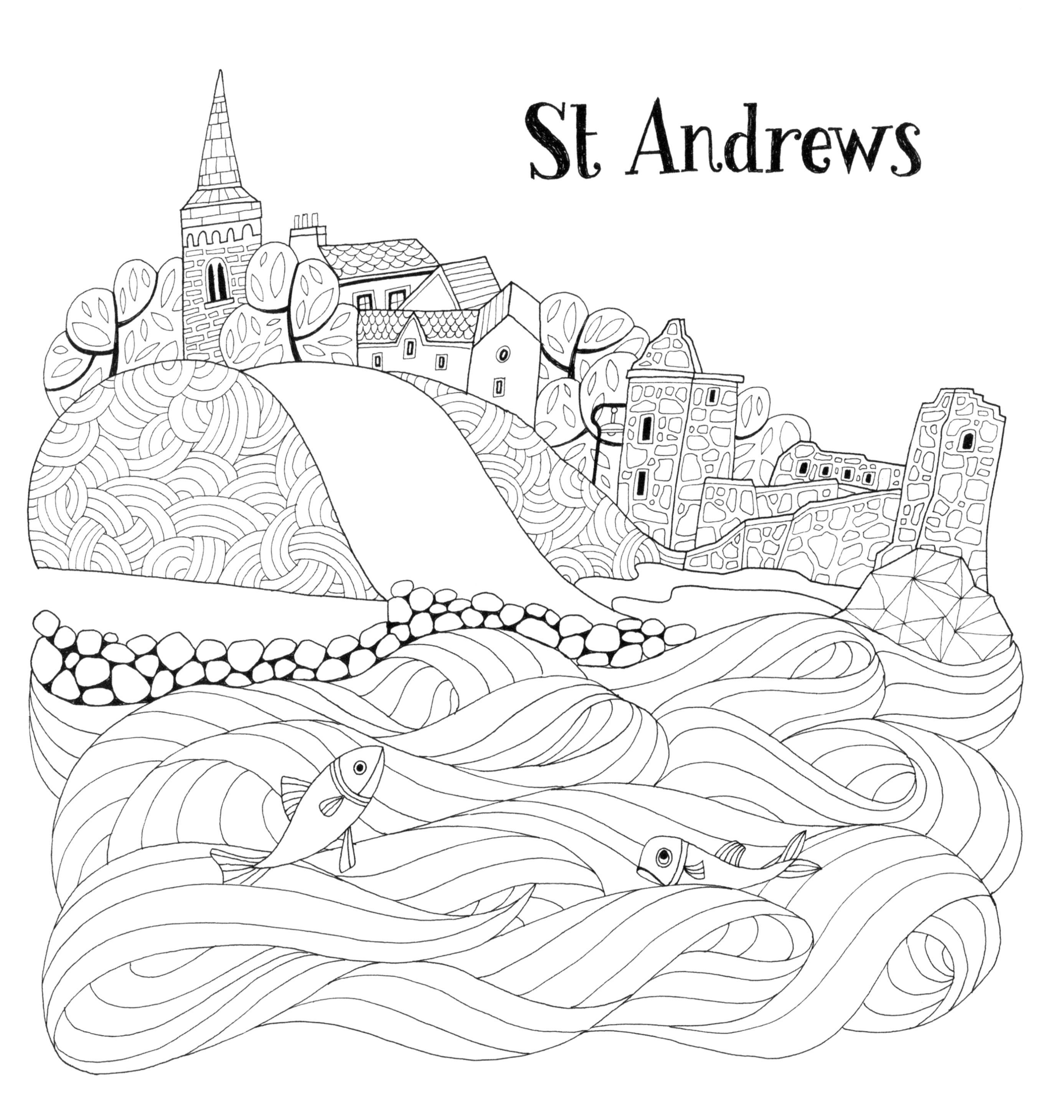
St Andrews

# Dunnottar Castle

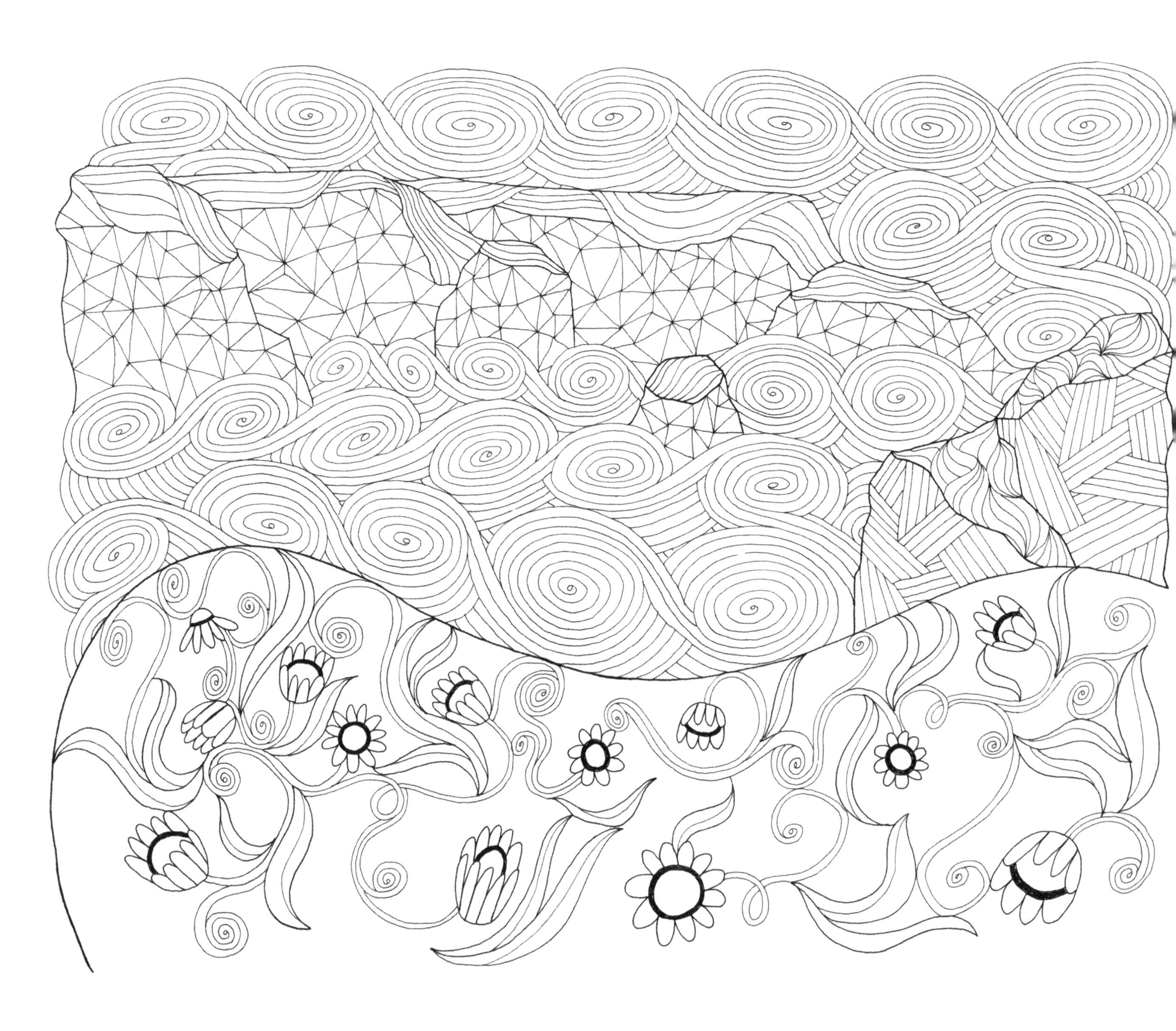

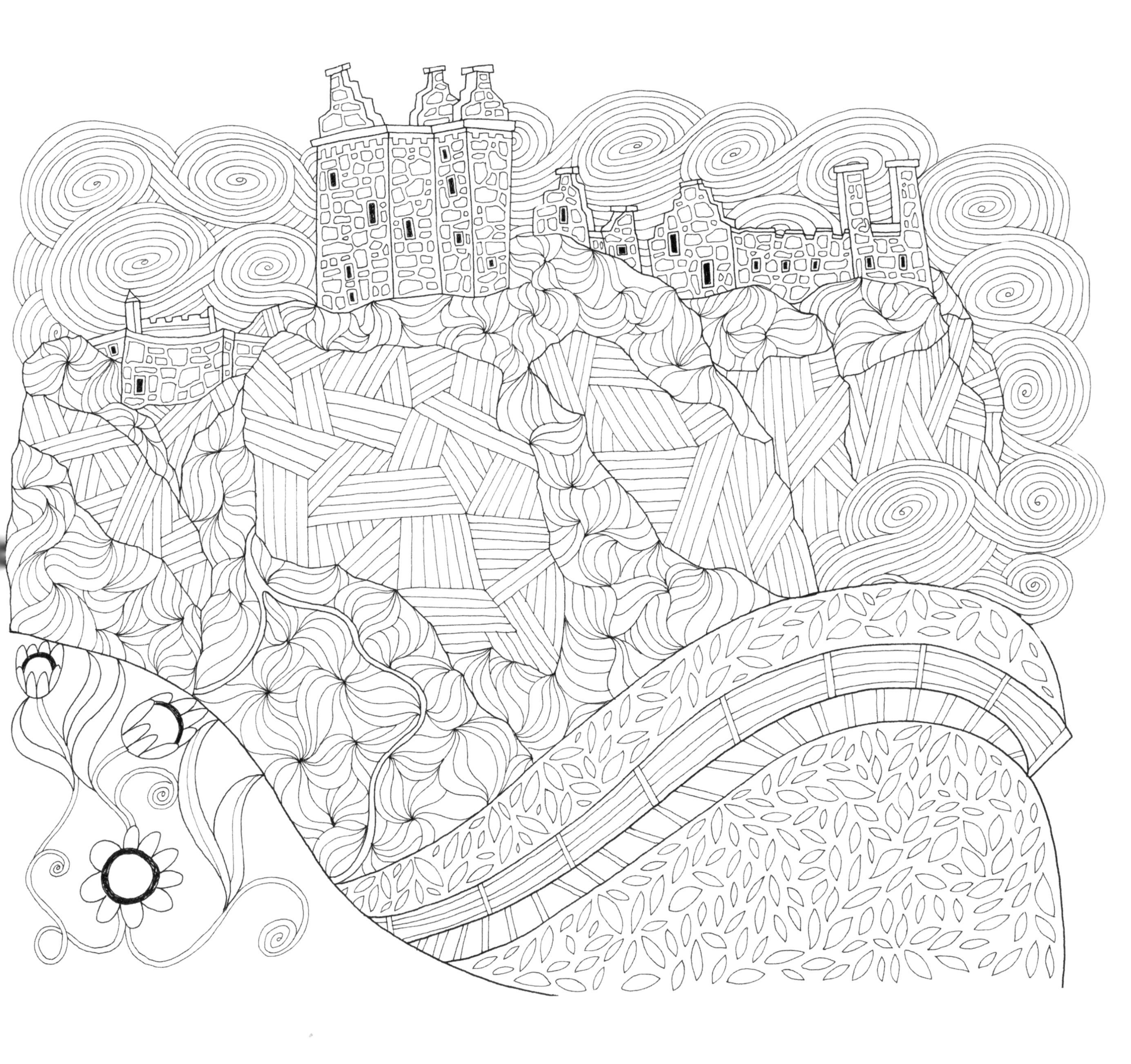

# Moray Firth Coast

Duncansby
Head

John o'Groats
JOHN O'GROATS
NEW YORK
LAND'S END
ORKNEY · SHETLAND
EDINBURGH

North
Ronaldsay,
Orkney

# The Old Man of Hoy

# Shetland Puffins and Friends

# Ullapool

# St Kilda

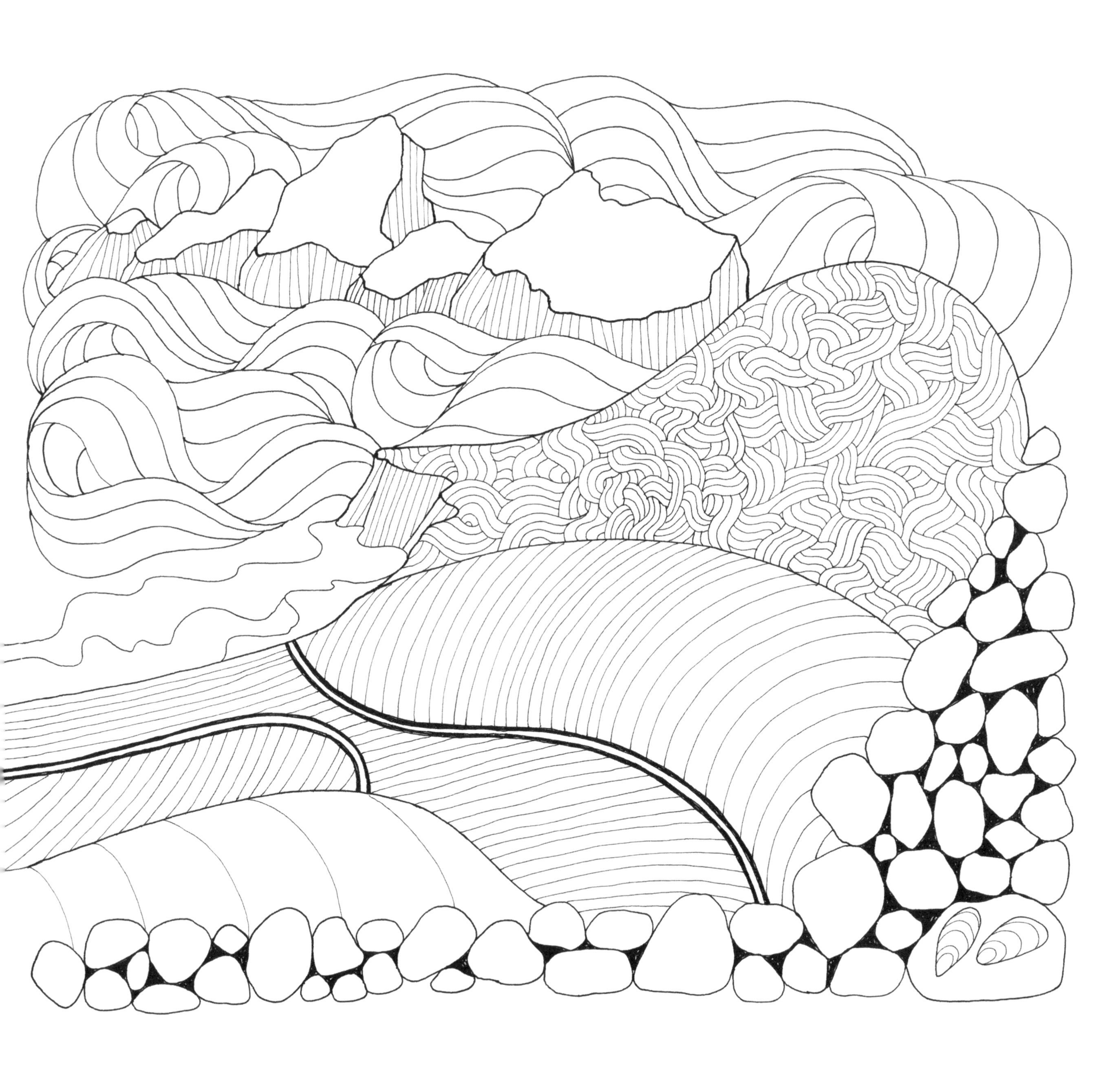

# Luskentyre, Harris

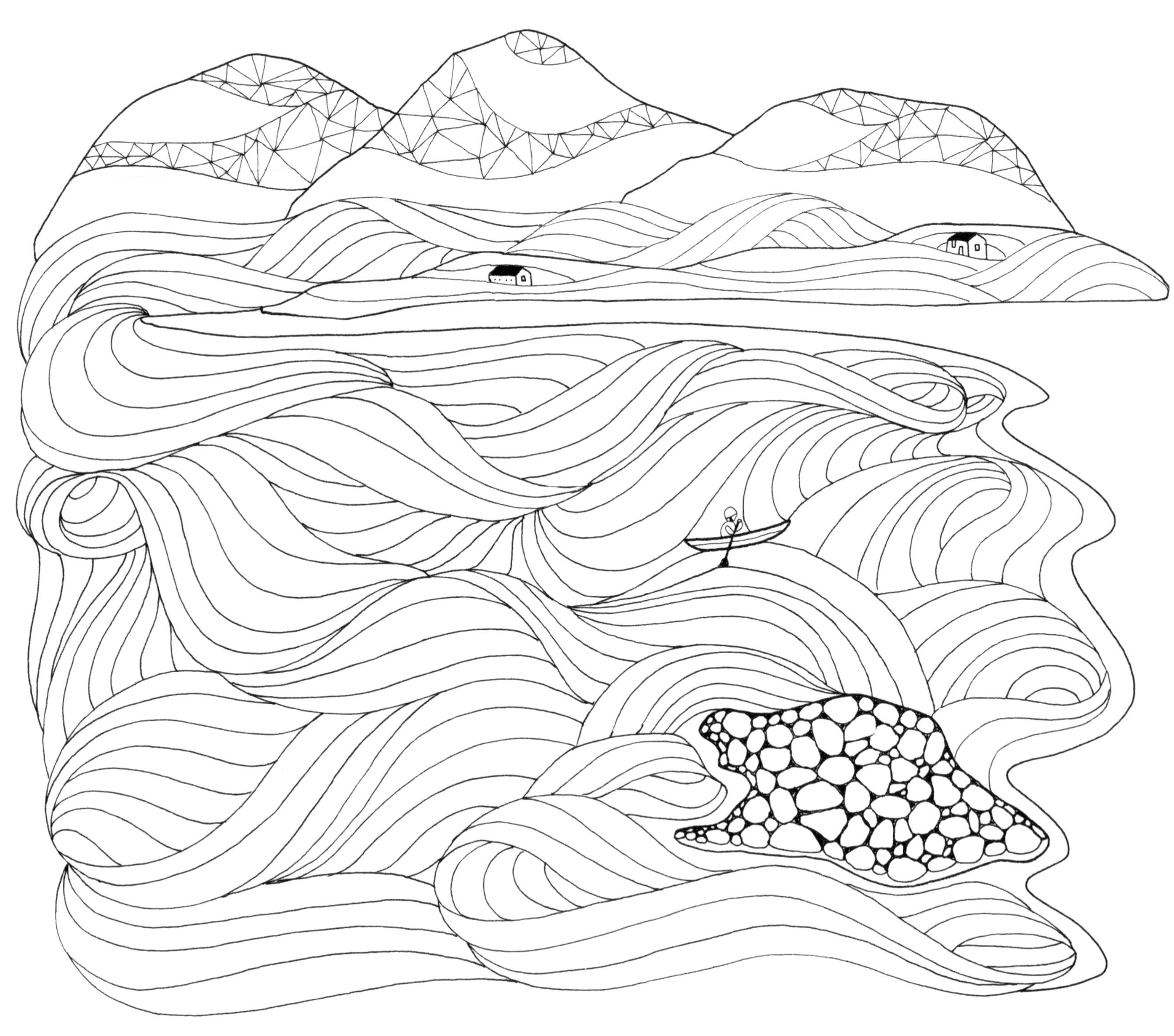

# Plockton

Fingal's Cave,
Staffa

# Castlebay, Barra

Windsurfing,
Tiree

Corryvreckan
Whirlpool

# PS Waverley

Culzean Castle

# Ailsa Craig

# Robin Rigg Wind Farm

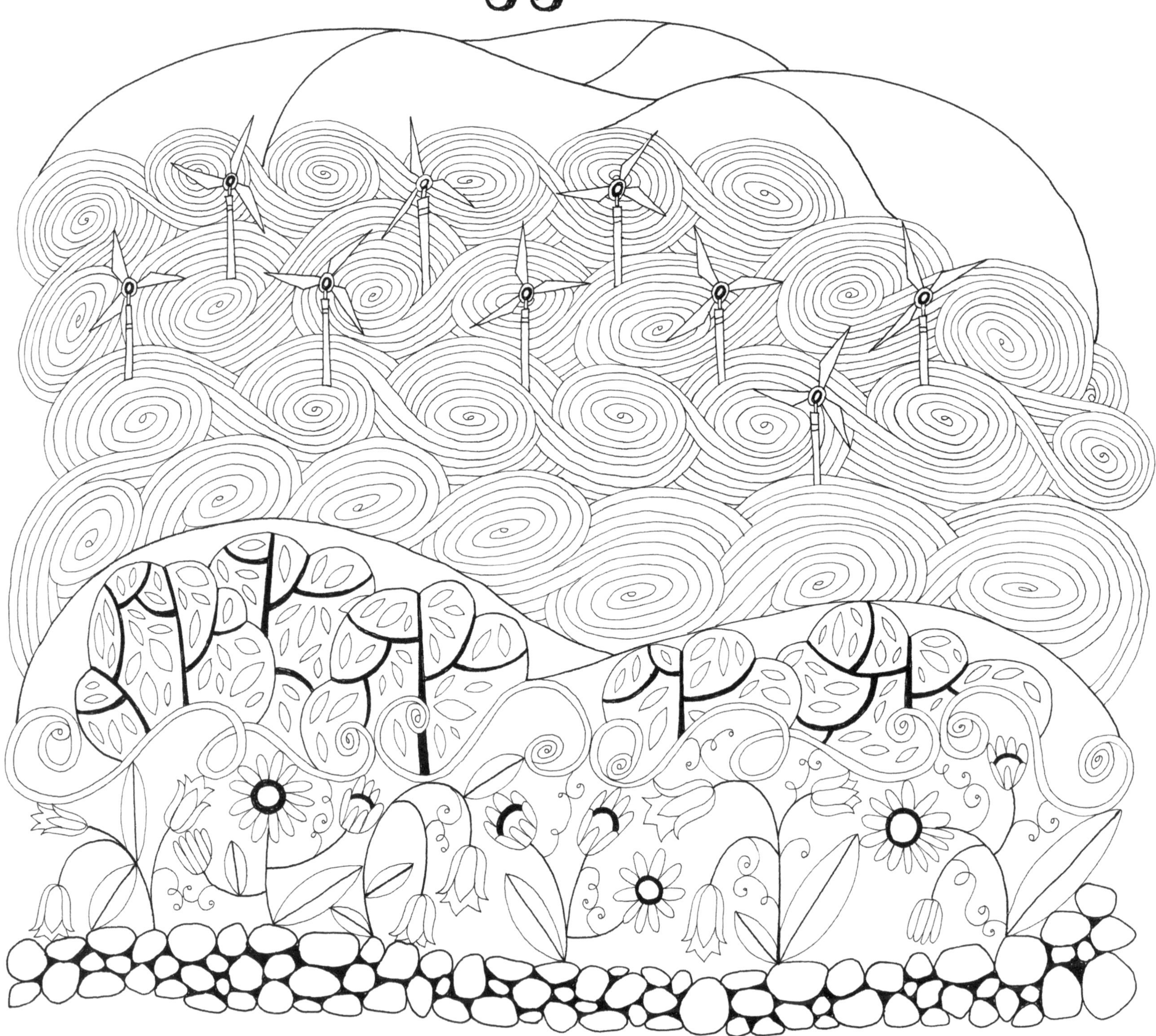

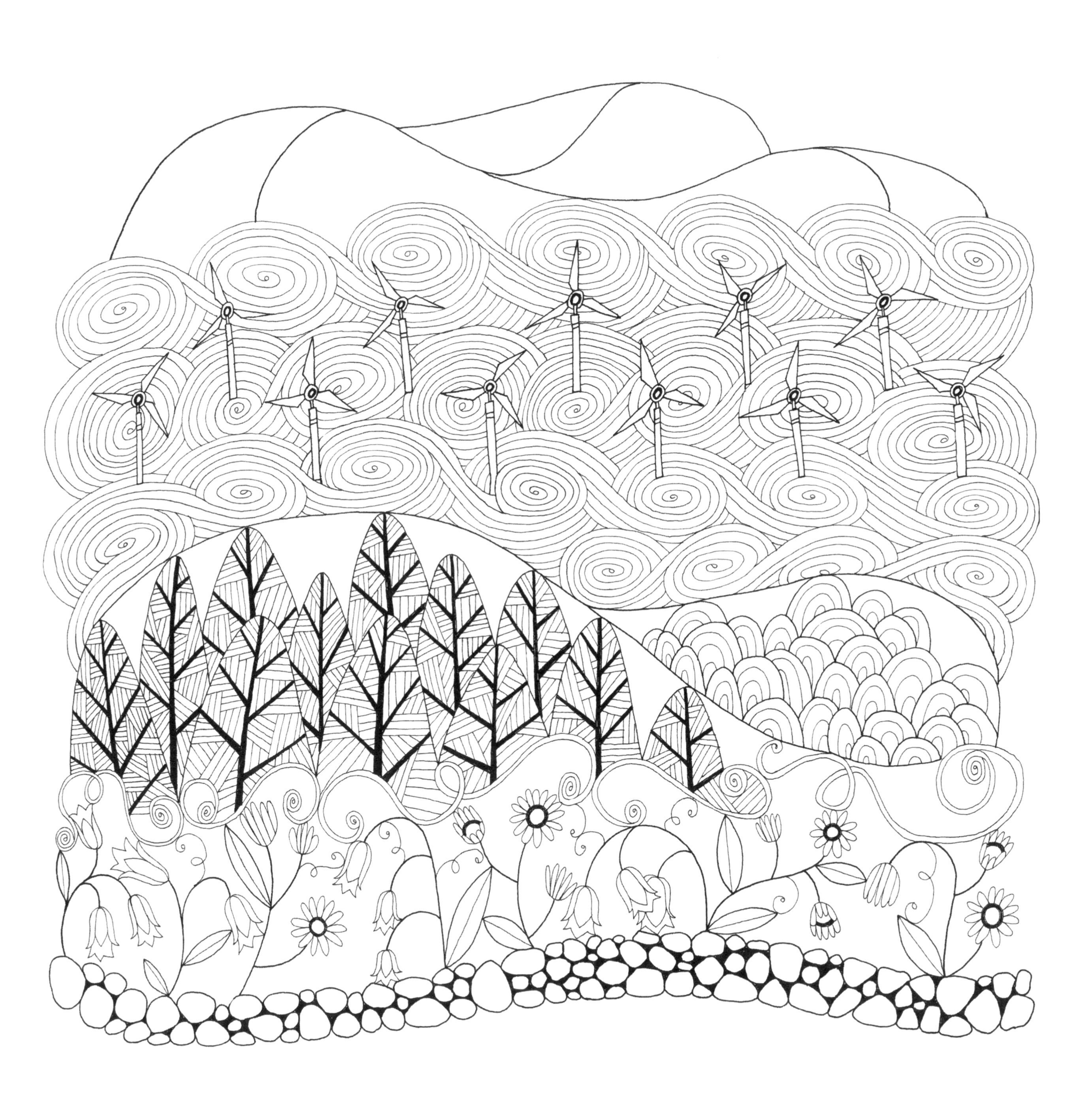

Rock Pools

The SCOTTISH COASTAL Colouring Book Journey
One: St Abbs
Two: Tantallon Castle and Bass Rock
Three: Culross
Four: Crail Harbour
Five: St Andrews
Six: Dunottar Castle
Seven: Moray Firth Coast
Eight: Duncansby Head and John o' Groats
Nine: Orkney
Ten: Shetland
Eleven: Ullapool
Twelve: St Kilda
Thirteen: Luskentyre, Harris
Fourteen: Portree, Skye
Fifteen: Plockton
Sixteen: Fingal's Cave
Seventeen: Castlebay, Barra
Eighteen: Windsurfing, Tiree
Nineteen: Corryvreckan Whirlpool
Twenty: PS Waverley
Twenty-One: Culzean Castle and Ailsa Craig
Twenty-Two: Robin Rigg WindFarm
Ten
Nine
Eight
Twelve
Thirteen
Eleven
Fourteen
Seven
Fifteen
Six
Seventeen
Eighteen
Sixteen
Five
Four
Three
Two
Nineteen
One
Twenty
Twenty-One
Twenty-Two